THE LIBRARY OF
FUTURE ENERGY

HYDROGEN POWER OF THE FUTURE

NEW WAYS OF TURNING FUEL CELLS INTO ENERGY

CHRIS HAYHURST

THE ROSEN PUBLISHING GROUP, INC.
NEW YORK

Published in 2003 by The Rosen Publishing Group, Inc.
29 East 21st Street, New York, NY 10010

First Edition

Library of Congress Cataloging-in-Publication Data

Hayhurst, Chris.
Hydrogen power : new ways of turning fuel cells into energy / by Chris Hayhurst.— 1st ed.
p. cm. — (The library of future energy)
Summary: Discusses the pros and cons of using hydrogen power to help fight air pollution and meet our growing demand for electricity.
ISBN 0-8239-3666-X (library binding)
1. Hydrogen as fuel—Juvenile literature. [1. Hydrogen as fuel. 2. Power resources.] I. Title. II. Series.
TP359.H8 H39 2003
333.793—dc21

2002002425

Manufactured in the United States of America

CONTENTS

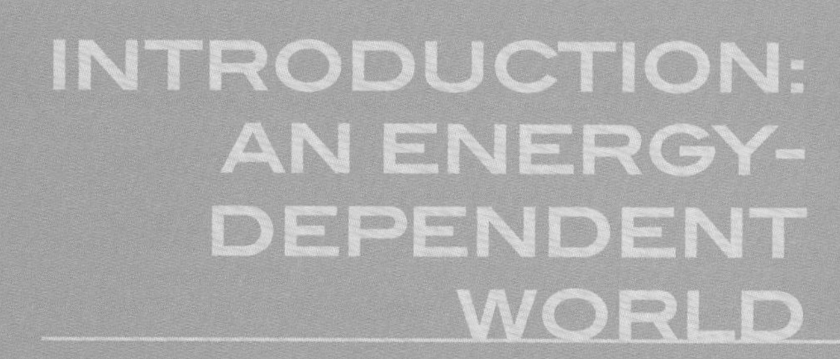

INTRODUCTION: AN ENERGY-DEPENDENT WORLD

Take a look around you. Everywhere you look—inside your house, outdoors on the street, up in the sky—the world is buzzing with activity. Airplanes fly through the air thousands of feet above the ground, zipping passengers between states, across countries, and around the globe. Cars cruise out of driveways, along alleyways, and down freeways. Homes, gyms, offices, and skyscrapers are lit up like Christmas trees so that the people within them can see, work, and play around the clock. Stereos blast music. Millions of computers hum. Televisions and movie screens glow with the images of actors, sports stars, and cartoon characters.

There's no doubt about it—the world is alive with technology. It's so alive, in fact, that we take this technology for granted. Think about it. Have you ever wondered why your computer turns on every time you press the power button? And have you ever even considered what enables the lights in your house to snap to life with a mere flick of a switch?

If you have the time to ponder such mysteries, you've probably come across the following word: energy. The dictionary defines energy as "the capacity for doing work." In other words, when an object is supplied with energy, it's able to work; without energy, it's useless.

So where does energy come from? Plants take and store energy directly from the sun. Animals, including people, obtain energy by eating food. The energy that powers technology, such as the electricity necessary to turn on a radio or the fuel required by an automobile engine, are generated from the earth's natural resources.

Energy experts like to divide energy generated from the earth's natural resources into two major categories: renewable energy, energy sources that are constantly replenished and never run out; and nonrenewable energy, energy sources that are depleted as they are used.

Renewable energy sources include solar and wind energy. You've probably seen solar panels on rooftops. These high-tech devices collect energy from the sun and convert it into electricity. Windmills do the same by harnessing the power of wind. Nonrenewable energy

As nonrenewable energy sources get used up, scientists are trying to tap renewable sources such as wind energy. Coal mines such as the one shown here *(left)* not only get depleted, they pose serious environmental hazards, unlike wind turbines *(right)*.

sources include coal, oil, and natural gas. Also known as fossil fuels, these materials are formed in the earth from the remains of ancient plants and animals.

Most of the energy used by people today comes from nonrenewable sources. There are many reasons for this, some political, some practical, and some economic. In recent years, however, it has become more and more evident that the future holds no place for nonrenewable energy. The reasons are both simple and complex. The simple part is this: At some point the world's supply of nonrenewable energy sources will dwindle down to almost

FOSSIL FUELS AND INCREASED EMISSIONS

According to the U.S. Environmental Protection Agency (EPA), fossil fuels burned to run cars and trucks, heat homes and businesses, and power factories are responsible for:

- 98 percent of U.S. carbon dioxide emissions
- 24 percent of methane emissions
- 18 percent of nitrous oxide emissions

nothing—or at least to levels that make them too expensive to use. As for the complex part, consider the environment: The burning of fossil fuels sends enormous amounts of pollutants into the air—air that humans, other animals, and plants rely on for life. Meanwhile, the use of renewable energy sources such as solar and wind power causes little or no harm to the environment. Why is this complex? Because different people put different values on environmental health. Some feel that a certain amount of pollution is just a fact of life. Others believe we should do everything we can to keep the earth as clean as possible. And still others fall somewhere in the middle.

Regardless, the reality is that we all need energy to survive—and not just the energy we get from the food we eat. We need energy to run the technological gizmos we've come to rely on in everyday life. You see, energy, is critical to society. And in the future, as the world's population continues to grow and our energy needs

FACTS

- American drivers fill their vehicles with nearly 100 billion gallons of gasoline every year.
- According to the U.S. Environmental Protection Agency, vehicle emissions currently cause nearly 60 percent of urban air pollution.
- If all vehicles were to run on hydrogen obtained from water, combined with a backup system fueled by solar or wind power, they would produce no pollution whatsoever.

expand worldwide, its importance will be magnified. The need to move toward renewable energy is obvious. Our lives depend on it.

But first, there's a problem. While renewable energy is abundant in most parts of the world, many of these sources are not consistent enough to supply our energy needs all the time. In Alaska and northern Canada, for example, winter brings darkness for months at a time and the sun barely makes it over the horizon. In places like these, solar energy is just not an option, at least not for most of the year. The only way renewable energy sources can become major power suppliers in such areas, and other places with similar limitations, is through a reliable and efficient means of storing and transporting that energy. A way must be found to harness renewable energy from one area where it's abundant, store it intact, and move it to an area that needs it. So what's the solution? Energy experts say the answer is hydrogen.

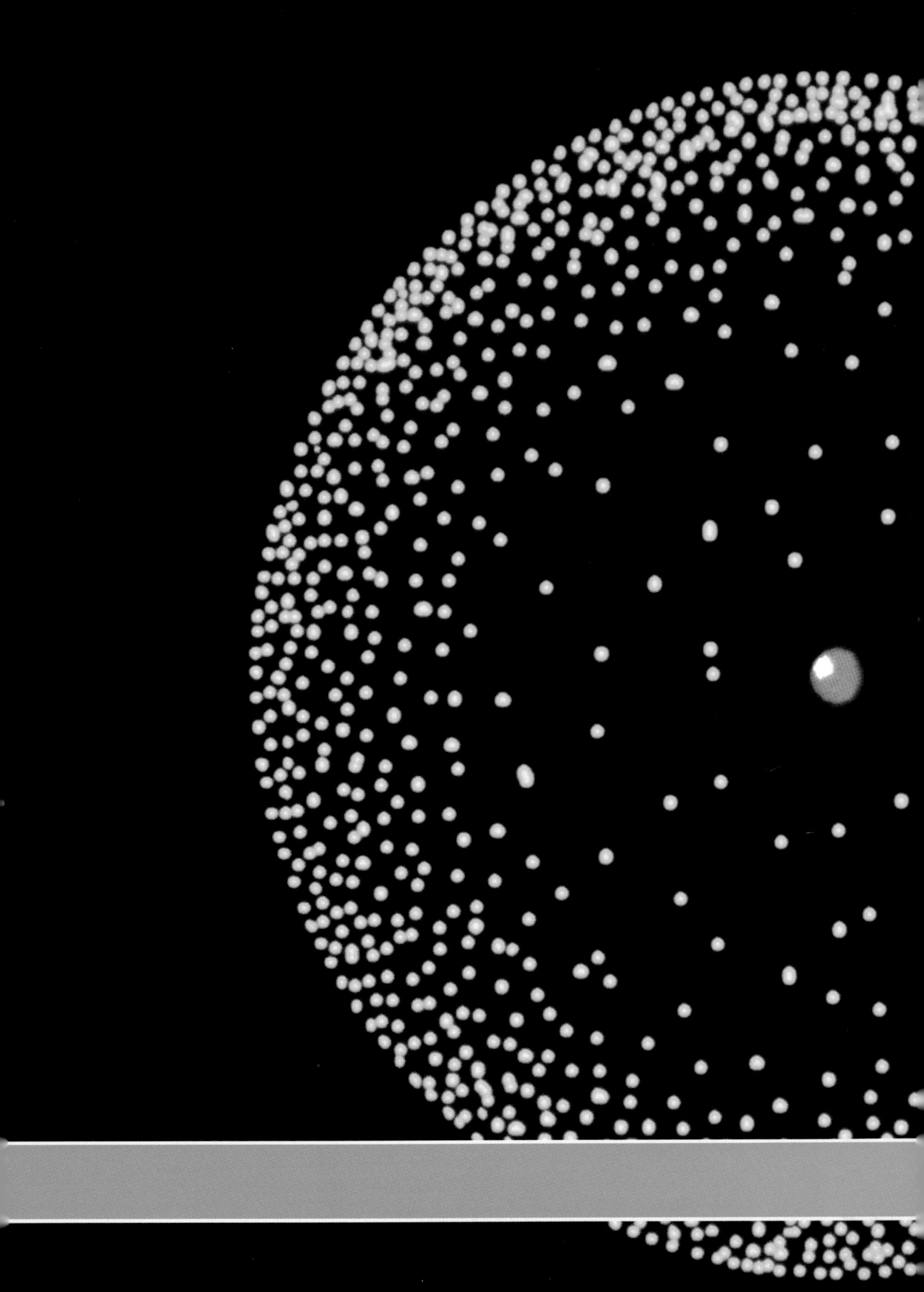

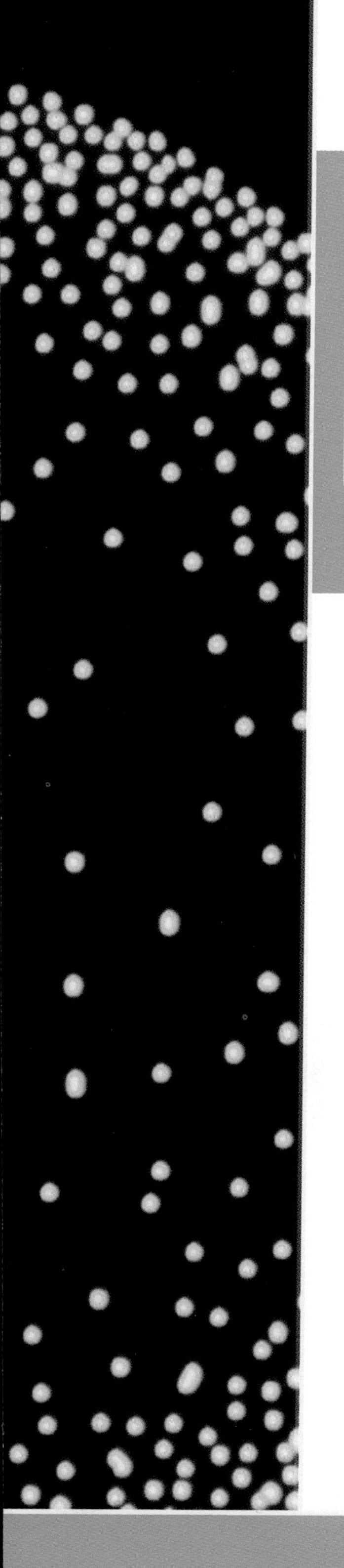

1 THE BASICS OF HYDROGEN

Hydrogen is an element—a substance that cannot be broken down into other substances through chemical reactions. Other elements you may have heard of include gold, silver, and oxygen. There are ninety-two naturally occurring elements known to scientists. What makes hydrogen unique is that it's the simplest and most abundant element of them all.

The symbol for the element hydrogen is H. Hydrogen is chemically very active—so active, in fact, that it almost always wants to get together with other elements. This getting together, so to speak, is known as bonding. When hydrogen bonds with other elements, it forms a compound, a substance

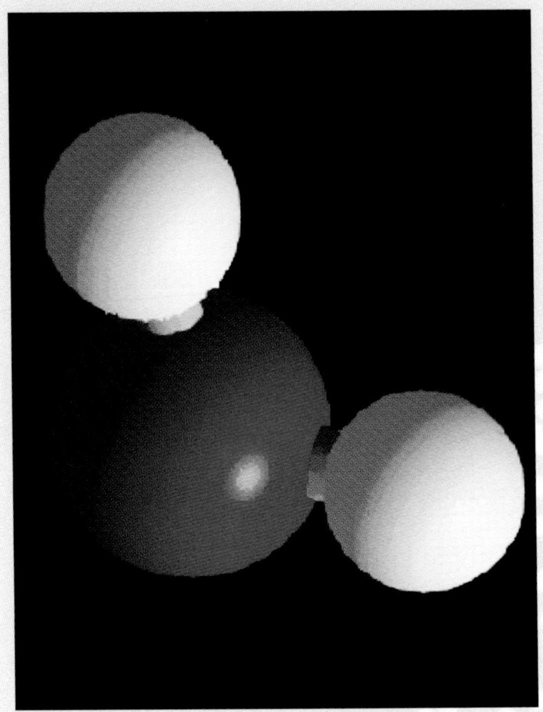
This illustration depicts a water molecule. The two smaller white spheres are hydrogen atoms attached to the larger red sphere, which represents an oxygen atom.

which combines two or more of the same or different elements.

Most of the hydrogen in the universe—and there's a lot of it—is found as hydrogen gas. The symbol for hydrogen gas is H_2. This is because it consists of two hydrogen atoms bonded together. Hydrogen gas is the most plentiful gas in the universe. By some estimates it accounts for 75 percent of the universe's mass. Still, you won't see it everywhere you look. In fact, hydrogen is invisible. It's also colorless, odorless, and tasteless.

Despite its abundance in the universe, hydrogen is not found as a gas here on Earth. Instead, in nature at least, it exists in compounds with other elements. The most common hydrogen-containing compound found on Earth is water, or H_2O. As the symbol suggests, water is two parts hydrogen and one part oxygen. Another compound containing hydrogen is methane. The symbol for methane, which is a gas, is CH_4. Methane is four parts hydrogen and one part carbon. There are many other compounds containing hydrogen as well. Hydrogen can be found in every living organism on earth.

Getting to the Hydrogen

The fact that earthbound hydrogen only exists in compounds bonded to other elements presents a tricky problem for anyone hoping to get their hands on a hydrogen molecule. For hydrogen to be useful as a source of energy, it must be separated from its feedstock—plucked away from its partners. The bottom line: Bonds must be broken.

For a professional scientist, breaking bonds is not a big deal. It's done in the laboratory all the time. The great thing about hydrogen is that its bonds are relatively easy to break—easier than, say, the bonds that hold together octane, the prime component of gasoline.

There are a number of ways scientists can go about the hydrogen-production process. Not surprisingly, a scientist makes the choice based on cost. If a process is expensive, it doesn't get much attention. On the other hand, if a process is fairly affordable, that's the one that is chosen. Likewise, if it takes more energy to produce the hydrogen than the amount of energy the hydrogen will ultimately provide, the process is not useful to industry. Two of the most common hydrogen-production methods used today are natural gas steam reforming and water electrolysis.

Natural Gas Steam Reforming

The most popular method of extracting hydrogen from hydrogen-containing compounds is through a process known as steam reforming. Relatively inexpensive, steam reforming involves two basic steps. First, natural gas (the same sort as the natural gas many homes have piped in for heating needs) is exposed to extremely hot steam which is at least 392 degrees Fahrenheit. When the steam—which is composed of hydrogen and oxygen molecules—meets the natural gas, a chemical reaction takes place. This reaction produces hydrogen as well as carbon monoxide and carbon dioxide. In step two, additional hydrogen is produced when the carbon monoxide is exposed to more steam. The end result is true to the process's name: The steam is "re-formed" so that its hydrogen and oxygen molecules have been separated from each other, and more hydrogen has been created.

Water Electrolysis

The other major hydrogen-extraction method currently in use is known as water electrolysis. In water electrolysis, electricity is used to "split" water into hydrogen and oxygen gas. An electric current is sent through a container of water. The current wreaks havoc on the bonds between the oxygen molecules and the hydrogen molecules, causing them to break. The oxygens bond

This photo illustrates electrolysis of water. The current from the battery splits the water molecules into their component parts, hydrogen and oxygen, which are seen as gas in the test tubes.

together to form oxygen gas, and the hydrogens are ready for the taking.

The biggest variation in the water electrolysis process is the source and the cost of the electricity used for the splitting. If the electricity is cheap, the splitting process is cheap. If the electricity is expensive, as is often the case, splitting water through electrolysis is impractical.

While a variety of energy sources can be used to provide the electricity for this process, burning fossil fuels like coal to make

HYDROGEN SAFETY

Before hydrogen can become a mainstream source of energy, it must first clear one major obstacle—the perception that it's dangerous. Thanks to a number of well-publicized major catastrophes, such as the explosion of the *Hindenburg* (see page 25) in which hydrogen was involved and people lost their lives, there is a widespread fear that hydrogen is not safe. The truth is, hydrogen, when handled correctly, is perfectly harmless and is even less dangerous than gasoline.

electricity pollutes the environment. The most promising source of electricity for this process should be clean and renewable like solar power, wind power, and hydropower.

STORING HYDROGEN

One of the most interesting and important things about hydrogen is that once it is stripped away from its parent compound, it can be stored and transported from where it was manufactured to wherever it is needed. The following storage techniques are currently being used and refined.

COMPRESSED GAS

One of the most popular ways to store hydrogen is as a gas. Storing hydrogen as a gas requires less energy than does storing it as a liquid. But hydrogen gas takes up a lot of space, so to store a significant

Hydrogen in its natural form is a gas. To make it easier to use and transport, it is usually compressed into its liquid form and stored in pressurized containers. In this photo, the space shuttle *Discovery* sits on its launch pad near a liquid hydrogen storage tank.

amount in a limited space, first it must be highly pressurized. The high-pressure gas can be stored in high-strength metal storage tanks. In the future, large quantities of hydrogen gas may be stored in specially designated holding areas like caverns or mines. It could then be directed through a piping system into homes and buildings where it is needed for energy, much in the same way natural gas is transported to residences today.

Liquid Hydrogen

Gases tend to take up a lot of space. Liquids, on the other hand, are far more dense than gases. This means that pound for pound, more liquid can be stored in a given amount of space than gas. Using special cooling techniques to convert hydrogen gas into liquid, scientists are able to store and transport much more of the fuel than would otherwise be possible. This is a tricky and inefficient process, because the hydrogen must be cooled down to negative 423 degrees Fahrenheit. This requires a large amount of energy. Nevertheless, liquid hydrogen may be the fuel of choice for future airplane fleets, as its compactness makes it ideal for transporting on aircraft with limited space.

Chemical Hydrides

Hydrides are chemical compounds consisting of hydrogen and certain metals like magnesium, nickel, copper, or iron. When hydrides are heated, they decompose and the hydrogen is released. One problem with hydrides, however, is they weigh a lot in comparison to the amount of energy they carry. Scientists are trying to develop hydrides that carry more hydrogen and more energy.

Getting to the Energy

Once hydrogen is split off from its original compound and stored, it's ready to provide energy. The energy that went into the splitting is carried in the bonds of the stored hydrogen.

THE REAL PRICE OF FOSSIL FUELS

Today the vast majority of the industrialized world's energy is supplied by fossil fuels. As the world's biggest energy consumer, the United States gets more than 80 percent of its energy from coal, oil, and natural gas.

Relying on fossil fuels has come at a price. According to the Environmental Protection Agency, atmospheric concentrations of carbon dioxide, which is released into the air with the burning of fossil fuels, have increased nearly 30 percent since the late 1800s. Concentrations of methane, another polluting product of industrialization, have more than doubled. The concentration of nitrous oxide has increased 15 percent. These pollutants are prime causes of the thick haze called smog that hangs over many cities. Smog can lead to asthma and other respiratory illnesses.

Most people don't consider the real cost of fossil fuels. If they did, they would see that the use of fossil fuels has increased health costs for treatment of diseases caused by air pollution. It is expensive to clean up oil spills and leaks from storage tanks, as well as to fight the effects of global warming and acid rain. When the real costs are factored into the equation, fossil fuels don't seem so cheap.

To get to this energy, the single bonds between hydrogen molecules must be broken. There are two ways this can be done. The first is through controlled burning of the hydrogen as fuel. Hydrogen fuel can be burned as either a liquid or a gas. The burning fuel generates heat which can be used to run a furnace

Fuel cells are a kind of battery which convert hydrogen and oxygen into electricity. Here, Bruce R. Rauhe (*left*), technical director of the Fuel Cell Center at Houston Advanced Research Center, talks about the fuel cells on display, with William Taylor *(center)*, director of the Texas State Energy Conservation Office, and Malcolm Jacobson (*right*), president of Fuel Cells Texas.

or power cars or airplanes. An engine that burns hydrogen produces virtually no pollution. The second way to get to the energy is through a fuel cell, which can generate electricity.

What Is a Fuel Cell?

Fuel cells are a lot like batteries. They produce electricity by combining stored hydrogen with oxygen in a process exactly opposite that of electrolysis. When oxygen and hydrogen are combined, they produce electricity, heat, and clean, drinkable water. Fuel cells are not like normal engines. They are completely stationary—that is, there are no moving parts. They're also perfectly quiet and very efficient. Most important, fuel cells don't produce any pollution.

Fuel cells can be used for many different purposes. The electrical energy they produce is good for anything that needs normal electricity—a television, a computer, or a light bulb. Fuel cells can also power cars. And when they're not being used, the energy stored inside the fuel cell stays put until it's needed. Unlike a regular battery, as long as there is hydrogen gas supplied to a fuel cell, it never needs to be recharged.

2 HISTORY AND DEVELOPMENT

First, take hydrogen the element. Since the day the universe came into being, it's been the key ingredient in the soup of existence. With so much hydrogen in stars, it is the most abundant element in the universe.

Still, it was a very long time—eons, in fact—before hydrogen was "discovered." That honor goes to a man named Henry Cavendish, an English chemist and physicist. In 1766, Cavendish realized that hydrogen was a separate substance, something that could be isolated from all other matter. Later, in the 1770s, Cavendish became the first chemist to make water from hydrogen and oxygen. He did this by igniting a mixture of these gases with an electric spark.

Henry Cavendish (1731-1810), an English chemist, physicist, and natural philosopher, was the first to discover hydrogen.

Over the next century, scientists in different parts of the world conducted more experiments with hydrogen and oxygen.

In London in the late 1830s, a man named William Grove invented the fuel cell. Also beginning in the early 1800s, a gaseous product called "town gas," made from coal, was used to supply light and heat throughout Europe and the United States. Town gas, which even today is still used in some places, was a mix of hydrogen, methane, carbon dioxide, and carbon monoxide gases.

The end of the town gas era came in the mid-1900s when natural gas fields were discovered and tapped. Town gas fell out of favor as pipelines made it quite easy and inexpensive to shuttle natural gas to homes and businesses.

Fuel cells came back on the scene in 1959 when a British engineer named Francis Bacon developed a fuel-cell system that would later be used by the National Aeronautics and Space

THE MYTH OF THE *HINDENBERG*

You may have heard about the *Hindenberg* disaster of 1937. The tragic event involved a hydrogen-filled airship known as a dirigible—a blimp-like craft powered by the controlled burning of hydrogen gas. As the *Hindenberg* maneuvered to its landing dock in Lakehurst, New Jersey (following an uneventful journey across the Atlantic from Germany), it caught fire. Thirty-six people were killed.

Contrary to popular belief, the initial fire was not caused by the hydrogen the *Hindenberg* carried for fuel. In fact, investigators have shown that the craft ignited as a result of stormy weather that created electrical charges in the atmosphere. The ship's outer surface material had been painted with a substance containing the same ingredients as rocket fuel, and when this substance came in contact with the electrical charges, it caught fire.

The hydrogen gas did wind up burning when flames from the surface spread to other parts of the craft, but that fire did not injure any of the passengers. Instead, most of those who perished did so when they jumped from the ship in a panic before it touched ground.

Some historians believe it was an act of sabotage.

British scientist Francis T. Bacon, a pioneer in the field of hydrogen energy, demonstrates, in 1969, the revolutionary cell he developed which generates electrical power by reversing the chemical process of storage batteries. The cell produces electricity by combining hydrogen and oxygen.

Administration (NASA) and the American space program. Then, later that same year, the first successful fuel-cell-powered vehicle—a twenty-horsepower tractor—was invented.

Also in the late 1950s, NASA was formed. NASA soon became the world's biggest consumer of liquid hydrogen, using the fuel in massive quantities to propel its aircraft and spaceships away from earth and high into the sky.

The 1950s can be considered the true start of the modern revolution in hydrogen-powered technology. Since then, scientists have

The space shuttle *Atlantis*, taking off in 1991. NASA is the biggest consumer of hydrogen, using it to power its space missions. This has given hydrogen the name rocket fuel.

GLOBAL WARMING AND THE GREENHOUSE EFFECT

The greenhouse effect refers to the heat-trapping abilities of the gases of the earth's atmosphere, especially carbon dioxide and methane. These gases function like the glass in a greenhouse, allowing some sunlight to pass down to the earth but then preventing some of the heat from returning to space when it rebounds off the earth's surface. Because this heat can't escape, as it is supposed to, it warms up the earth. This is global warming.

A certain amount of warming is essential for life, but too much can cause serious problems. According to the EPA, the average temperature at the surface of the earth has increased by about one degree Fahrenheit since the late 1800s. This may not seem like a lot, but many of the earth's organisms are very sensitive to temperature changes. EPA studies have also shown that the ten warmest years of the 1900s all occurred in its last two decades—between 1980 and 2000. This may indicate that the earth is warming more rapidly now than ever before. Other sure signs of global warming include a decrease in the amount of snow in the Northern Hemisphere; less floating ice in the Arctic Ocean; and a worldwide rise in sea level (as the ice melts, sea level rises). All of these changes can have serious consequences for plants and wildlife.

Scientists have shown that one of the major culprits in global warming is the unnatural increase in carbon dioxide levels. Carbon dioxide is one of the main gases released into the atmosphere with the burning of fossil fuels. For this reason, many countries around the world are increasingly turning to clean, carbon dioxide-free renewable energies as alternatives to energy from coal, oil, and gas.

GET ON THE BUS

According to studies by the New York City-based United Nations Development Program (www.undp.org), replacing the world's diesel buses with clean-burning fuel-cell-powered buses (running on hydrogen produced from natural gas) is a sure way to help improve global air quality. In fact, doing so in developing countries alone would reduce the amount of carbon dioxide released into the air by nearly 440 million tons per year.

made huge strides in developing more efficient ways to produce hydrogen, better ways to store it, and innovative ways to use it.

Today hydrogen is still used primarily to launch space vehicles, but it is also being used in many other industries. All of the major automobile manufacturers are working hard to get hydrogen-powered vehicles on the road in the next few years. Governments, universities, private companies, and research groups around the world are conducting extensive research and development into the promise of hydrogen. They are hoping to improve upon existing technology to bring hydrogen into widespread use.

OPEC

3 THE POLITICS OF ENERGY

Considering all the health and environmental benefits that come with the use of clean and renewable energy sources, one can't help but wonder why we continue to use fossil fuels like coal and oil. After all, why pollute the air with countless tons of carbon dioxide when abundant and nonpolluting fuels like hydrogen are just waiting to be used?

The answer is tricky. In an ideal world, coal would remain buried in the ground. Oil would never see the light of day. We'd drive our fuel-cell-powered cars, fly in hydrogen-powered planes, and spend long nights beneath hydrogen-fueled lights. However, because of politics and the reality of economics, the world of energy is not ideal.

The United States government, sponsoring research to further the uses of hydrogen power, especially in daily life, recently announced a partnership with private industries to develop hydrogen cars. Here Senator Carl Levin addresses the media as *(left to right)* Chrysler group president Dieter Zetsche, Energy Secretary Spencer Abraham, General Motors chairman Jack Smith, and Ford vice president Will Bodies look on.

Big businesses—like oil companies, for instance—have serious influence in the political world. They contribute money to political campaigns. They work to place pro-business politicians in key positions in the federal and state governments. They help people get elected, and once that is achieved, they expect their favors to be returned.

With that in mind, it's easy to see why conservation and use of alternative energy resources are often far down the list of priorities for many elected officials who make the United States heavily dependent on oil and coal—fossil fuels. As a result, the United

States continues to import oil from foreign countries, despite pressure from much of the public to make alternative energy a priority. And coal will continue to be burned at power plants, despite the fact that it's one of the planet's biggest polluters—and supplies are slowly dwindling.

A petroleum engineer oversees a section of an oil pipeline. The United States's dependence on foreign oil has spurred new attempts to find cheaper, safer, and more reliable alternatives.

Importing oil from foreign countries comes at a price. The United States now depends on the Organization of Petroleum Exporting Countries (OPEC) for more than 50 percent of its oil supply. Unfortunately, some of the countries that belong to OPEC have terrible human rights records and are known sponsors of terrorism. In an age when the United States has declared "war on terrorism," this presents a difficult double standard. On the one hand, the U.S. government says no to terrorism. But on the other hand, it asks countries supporting terrorism if they would sell us their oil (at an affordable price). When we buy oil from these countries, we are funding people and ideas we are fighting against.

ALL ABOARD FOR CLEANER AIR

An aerial view of smog hanging over the city of Los Angeles in 1988

One of the most frustrating problems in many of the world's largest cities is how to minimize the smog, or ground-level ozone, that pollutes the air. Ozone occurs naturally in the upper atmosphere that surrounds the earth and is essential for life. Acting like a vast, gaseous blanket, the ozone layer protects living organisms by filtering harmful ultraviolet rays from the sun before they can reach the earth's surface. But ground-level ozone is an entirely different story. This hazy, low-hovering, yellowish layer of air, visible most often in heavily populated areas on hot summer days, is harmful to life. High concentrations of this type of ozone can lead to severe breathing problems in children with asthma. It can affect elderly people suffering from respiratory illnesses and harm the lungs of healthy adults who are exercising outdoors.

The first step in the creation of ground-level ozone is the release of two specific types of gas compounds into the air—oxides of nitrogen and volatile organic compounds. These gases have a number of different sources. Nitrogen oxides are most often formed when fossil fuels are burned in motor vehicle engines and power plants. Volatile organic compounds (VOCs) come from things like auto exhaust, gasoline vapors, and certain chemical-based products like paint.

The second step in the formation of ground-level ozone is the reaction of these gases with heat and sunlight. Heat and sunlight cause chemical reactions that turn the volatile organic compounds and oxides of nitrogen into ozone. Ground-level ozone is worse in the summer because the higher temperatures and increased sunlight create more smog.

Smog is a major health issue in many urban areas of the United States. In big cities like Los Angeles and Salt Lake City, for example, health officials often issue warnings to the public about high levels of smog, often with stern recommendations for people to stay inside. Fortunately, most U.S. cities are finding ways to reduce smog. Through stricter emissions tests on motor vehicles and tighter controls on air pollutants from factories and power plants, cities are working hard to make breathing easier.

In many other parts of the world, progress toward cleaner air has not been so certain. Developing countries, which often have enough difficulty just keeping their economies running, are less likely to make environmentally sound restrictions a priority, especially if they believe those restrictions might prevent economic growth.

In six of the world's most populated and polluted cities, however, things are about to change for the better. Thanks to a new program sponsored by the New York City-based Global Environment Facility, Mexico City, São Paulo, Cairo, New Delhi, Shanghai, and

Beijing will receive fuel-cell-powered buses for use in their transportation systems. The intention is to help these cities meet the demands for better urban transport while helping to reduce their air pollution. Fuel-cell-powered vehicles release no VOCs or oxides of nitrogen into the air, where they could form smog. The only gas emitted from the tailpipe of such a vehicle is entirely clean and harmless—water vapor.

The Global Environment Facility hopes its $60 million program will help fuel-cell-powered buses become more popular in the developing countries that most need them. By demonstrating to the world's largest and most polluted cities that fuel-cell buses are both economical and a great way to reduce smog, the group hopes this transportation technology will eventually be accepted and used worldwide. If it is, one thing is for sure: The air we breathe will be far cleaner than it is today.

To solve the problem of oil shortage, some politicians have recommended that we pick up the pace of oil drilling on American soil, in places like the Arctic National Wildlife Refuge in Alaska. They say that by increasing the amount of oil we produce in the United States, we can decrease the amount we need to import. Unfortunately, this creates an entirely new problem. If

In its search for alternatives to foreign oil, the Bush administration has proposed drilling in parts of Alaska, including the National Arctic Wildlife Refuge. The refuge houses many endangered or dwindling species, such as the caribou *(above)*. Environmentalists are strenuously opposing the proposal.

we start drilling in pristine natural areas, we will hurt our environment and wildlife. While replacing some of the oil we import from foreign countries with oil from the United States would reduce our political problems, it would injure our air quality, water quality, and overall environmental health.

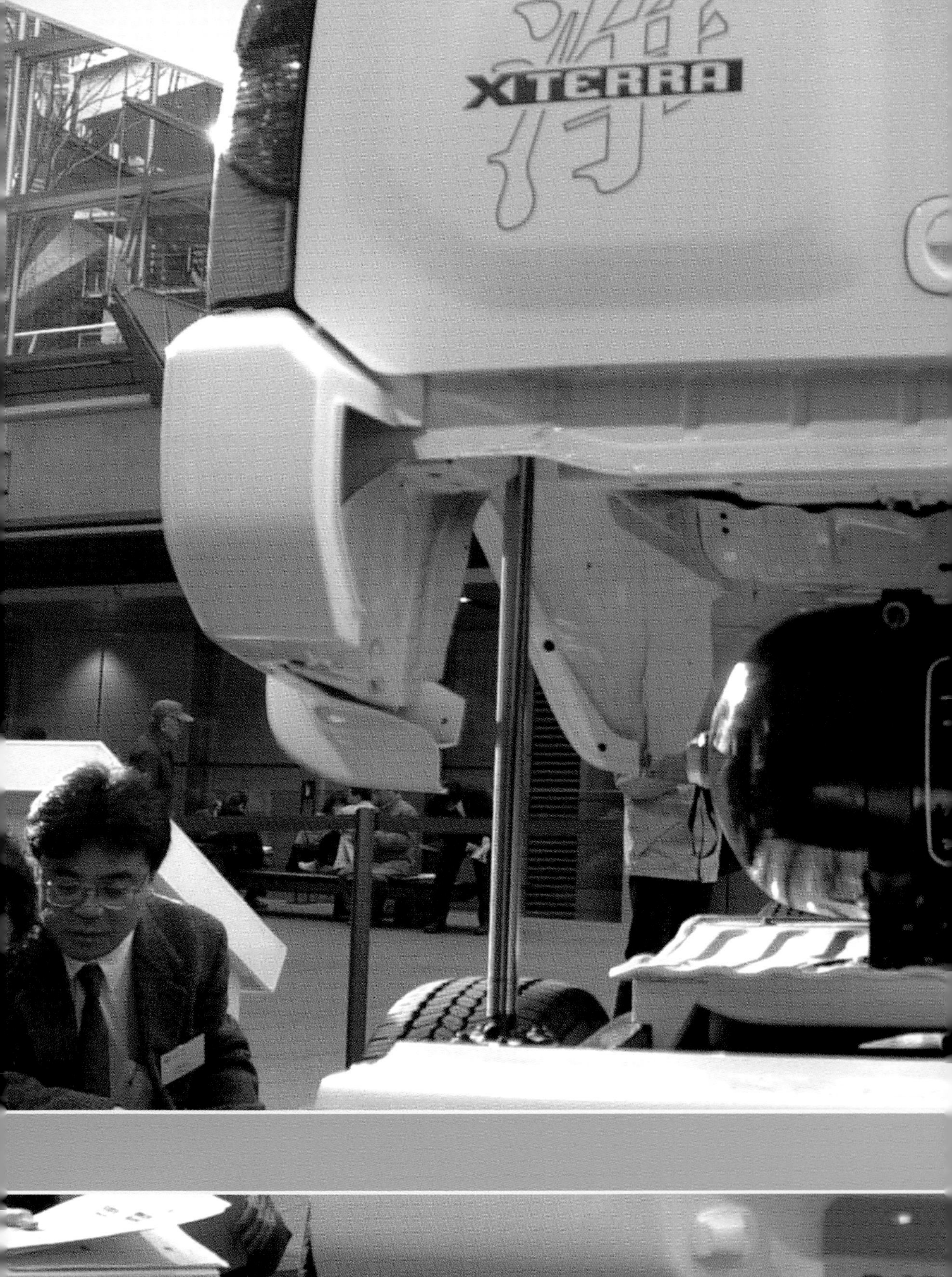
XTERRA

THE PROS AND CONS OF HYDROGEN

Hydrogen may hold the answer to the world's energy needs, but that doesn't mean it comes without drawbacks. There are pros and cons to using hydrogen just like there are pros and cons to using any form of energy or energy carrier. Consider both the good and the bad aspects of hydrogen and then decide for yourself if hydrogen has what it takes to solve the world's energy problems.

THE BENEFITS OF HYDROGEN

Hydrogen holds tremendous promise as an energy of the future. When it comes to the environment and political stability, hydrogen could be a tremendous help.

ENVIRONMENTAL BENEFITS

By far the biggest benefit of using hydrogen energy is that it is clean. When hydrogen is used for fuel, the only exhaust is water which can be recycled right back into the atmosphere. It can be resplit in the hydrogen-production process to isolate the hydrogen, and reused for fuel. The neverending cycle is pollution free!

According to the U.S. Department of Energy (DOE), the public will most likely use hydrogen on a large scale when it is added to transportation fuels. Mixing hydrogen with commonly used fuels like gasoline and natural gas, improves the efficiency of these fuels and reduces pollution. In fact, as the DOE points out, nitrogen oxide emissions from modern-day engines can be reduced 30 to 40 percent by adding a small amount of hydrogen to gasoline.

POWER POTENTIAL

According to the Department of Energy, hydrogen, the most abundant element in the universe, has the highest energy content per unit weight of any known fuel. It's incredibly lightweight and can be compacted into extremely small spaces. When hydrogen cools down, it changes from a gas to a liquid. Liquid hydrogen takes up just $^{1}/_{700}$ as much space as gaseous hydrogen! This makes hydrogen a great fuel for transportation, where small size and high efficiency are critical.

But wait, there's more. Not only is hydrogen abundant, high in energy, and easy to compact, it's also exceedingly simple. Hydrogen is two molecules stuck together by a single bond. Breaking this bond is a piece of cake, so hydrogen can release energy very quickly. This makes hydrogen a far more powerful fuel option than gasoline which has many more bonds and, therefore, releases energy much more slowly.

Unitrans, a bus system run by students in Davis, California, introduced buses powered by natural gases in 2001. One of the buses, the first in the country, used a mixture of natural gas and hydrogen.

Versatility

Taken alone, renewable energy sources are not exactly versatile. Sure, the sun can supply plenty of energy while it's shining, but after a week or two of stormy weather it suddenly doesn't look so promising. The energy collected by photovoltaic cells (the components of solar panels) is good while it's available, but when it's not . . . well, it's not.

AND STANDING IN THIS CORNER . . .

According to the Hydrogen Research and Applications Center of the Florida Solar Energy Center, hydrogen is a true heavyweight when it comes to power, but a lightweight when it comes to weight. Pound for pound, hydrogen can produce nearly three times the amount of energy as gasoline and seven times that of coal. This world-class weight-to-power ratio is one of the main reasons NASA uses hydrogen to launch its spaceships. It's also why hydrogen is the most likely candidate to replace jet fuel as the fuel of choice for jet airplanes. Who said bigger is better, anyway?

Hydrogen solves this problem. Hydrogen can store energy—in the form of highly pressurized gas or supercooled liquid, or in compact hydrides—and carry it quickly and efficiently wherever it's needed. In addition, shipping hydrogen long distances by pipeline costs just a fraction of sending electricity—another energy carrier—the same distance.

The ability to store and ship hydrogen solves two problems at the same time. Many sources of renewable energy are located far from where energy is needed. Imagine a wind farm in the middle of the lonely Wyoming plains, or a solar array gathering sunshine in the remote Arizona desert. Hydrogen offers the best, most efficient way to deliver energy from the source to the consumer, where it can be used to power so-called zero-emission vehicles, produce electricity, fuel aircraft, and heat homes and offices.

THE ROAD TO POLITICAL STABILITY

Hydrogen brings with it the promise of a much more politically stable world. Because hydrogen can be produced anywhere, no one country, leader, or region will ever be able to dictate its supply, price, or distribution. This is in stark contrast to oil, for example, which, available in just a few geographical regions, is subject to the changing political climate. If you consider all the wars that have been fought over oil, all the lives that have been lost as a result, and the people who have suffered the consequences, the implications are staggering.

THE PROBLEMS WITH HYDROGEN

The problems with hydrogen can more accurately be termed "challenges." For if hydrogen is to succeed as the answer to the world's energy wants and needs, many obstacles must be ducked, jumped, or destroyed first.

COST AND TECHNOLOGY

The reality of the situation is this: U.S. consumers are, for the most part, not driven by environmental concerns. They don't wake up in the morning and worry about the rainforests in Brazil or the air quality in Mexico, the water supply, acid rain, or global warming. What they do consider is price. When the cost of

A gas station in Connecticut tries to lure drivers with regular gasoline priced under a dollar. Many motorists regulate their oil consumption according to the current price of gasoline.

heating oil goes up, they notice. And when the cost of fuel at the local gas station goes down, they get right in there and fill up their tanks.

For now, hydrogen is too expensive to produce and use for it to compete with the cheap fossil fuels which are already on the market. Until the price of hydrogen goes down, consumers just won't bite.

Unfortunately, if manufacturers are afraid that consumers aren't interested, they have no incentive to offer hydrogen as an option. And if hydrogen is not produced on a large scale, the price will never go down.

What is encouraging, however, is that a hydrogen-based future is inevitable. Eventually fossil fuels will run low, and their prices will go up. Hydrogen will become a practical, affordable, and dependable alternative. Automobile industry leaders know this, and they're doing everything they can to prepare.

Infrastructure and Tradition

The world as we know it runs on oil and coal. That's the way it's been and, that's the way it's going to be. Or so it seems . . .

To understand just how big a deal it is to change the way the world gets its energy, consider the automobile industry. The entire automobile business—from engines to refueling stations—is built around oil-based fuels. There are thousands upon thousands of gas stations in the United States alone, all of which would have to convert their facilities to supply hydrogen instead of gas. This can't happen overnight.

Fortunately, hydrogen can be phased in gradually.

5 INTO THE FUTURE

The world today is much different than it was 100 years ago. A century ago, there were no computers and there was no Internet. Cell phones and satellites didn't exist. Spaceships were just a dream, and to suggest that jets would one day dominate the skies would have seemed crazy.

Just 50 years ago the world's population was 2.5 billion. Today that number has climbed well past 6 billion and is increasing at the rate of about 70 million people every year. The world is becoming more and more crowded, and its resources more and more strained.

One of the biggest problems facing the world today—a global environmental

Every major automaker is working to develop a hydrogen-powered automobile. Pictured above is BMW's 740hL hydrogen-powered sedan with a custom see-through hood, unveiled in Los Angeles in July 2001. BMW claims their luxury car cuts tailpipe emissions by 99.5 percent, blowing out nothing but water and steam. It has a 12-cylinder engine and can reach a speed of 141 miles per hour.

crisis—is a direct result of this massive population increase. We're not about to run out of energy—at least, not right away. But we need energy that doesn't pollute the air or water, that doesn't destroy the forests, and that doesn't lead to global warming. We need an environmentally friendly, renewable energy—the energy of the future.

For widespread renewable energy to become a reality, however, we must find ways to store and transport that energy. And this can be accomplished with hydrogen. But many things must happen

It Takes Money . . .

According to the U.S. Department of Energy, the federal government allots an average of about $18 million each year for hydrogen research and development. This money, plus funds provided by private organizations with an interest in developing hydrogen as the fuel of the future, is being used to support promising hydrogen research programs throughout the United States. One particularly interesting project is being staged at the Hydrogen Research and Applications Center of the Florida Solar Energy Center at the University of Central Florida. One of the program's intentions is to develop more efficient ways to split water into hydrogen and oxygen. Other research is being conducted in places like the Department of Energy's own National Renewable Energy Laboratory in Golden, Colorado; Texas A & M University in College Station, Texas; the Brookhaven National Laboratory in New York; and the Hawaii Natural Energy Institute in Honolulu, Hawaii.

first. Scientists are searching for ways to achieve this goal, but there is a lot of work to do, including the development of better technologies that allow hydrogen to be safely produced, stored, transported, and used.

Research is underway in the United States and elsewhere in the world. In Russia, for example, scientists have already developed and tested a high-tech airplane that is fueled by liquid hydrogen. NASA in the United States is doing the same, but they hope to take the technology a step further by using what is known as slush hydrogen instead of liquid hydrogen. Slush hydrogen is a

A VISION OF THE FUTURE

"To achieve our vision of cleaner, smaller, and more efficient sources of energy, we will also expand our exploration of the role of fuel cells and hybrid engines.

"Fuel cells, which can run on hydrogen, or traditional fuels that convert to hydrogen, offer the opportunity to address two different challenges. First, they may serve as the backbone of the distributed energy network. Second, as the auto manufacturers are already discovering, they offer the opportunity to change dramatically the debate about fuel efficiency.

"Hybrid vehicles, powered by traditional combustion engines and either batteries or fuel cells, already point toward a day when we can significantly curtail our reliance on foreign oil.

"Earlier this year, I glimpsed the future of fuel cells at DOE's Argonne National Lab. They are getting smaller, more powerful, and more useful virtually every day. In just the past four years, they've reduced the fuel processing system from the size of a minivan to the size of a driver's seat in a minivan. And further advances are certainly on the horizon. Seeing these fuel cells convinced me that our vision, which embraces the American commitment to a cleaner environment, provides a realistic path toward the use of energy in the future.

"For centuries, we have lived and prospered in a carbon-based economy. Fossil fuels powered ships, warmed homes, fueled automobiles, fired the revolution in flight, and the revolution in information technology. Energy sources like coal and oil once replaced an economy based on horsepower. So our carbon-based economy may itself pass from the scene to be replaced, perhaps, by hydrogen.

> "The president's plan directs us to explore the possibility of such an economy and such a future. The use of hydrogen, if realized, offers the possibility of completely clean energy—its only byproduct is water. And, since hydrogen is the most common element in the universe, it offers an essentially limitless source of energy."
>
> —From U.S. Energy Secretary Spencer Abraham's speech at the first annual Energy Efficiency Summit, October 25, 2001

mixture of liquid and solid hydrogen with a lower temperature and higher density than liquid hydrogen. The slush form of hydrogen would be more practical to use in the limited space available in aircraft.

Automobile manufacturers are also leading the way in research. Just about every major automaker in the world is working to develop a fuel-cell vehicle that can compete with the gas-powered cars and trucks already on the road. Manufacturers, realizing how important hydrogen's role will be in the future of transportation, are doing everything they can to avoid being left behind. Someday hydrogen-powered cars will be a common sight on roads and highways.

Other research is focused on hydrogen production, storage, and use. According to scientists at the Department of Energy's

National Renewable Energy Laboratory, the latest developments include the following.

Production Research

Scientists are working hard on photoconversion production. One method of photoconversion in the works relies on genetically engineered biological organisms to absorb sunlight, split water, and produce hydrogen. Producing hydrogen is natural for some organisms. Scientists are merely trying to figure out how they can best use these organisms to make hydrogen on a large scale.

Another photoconversion technology currently in development uses semiconductors to generate an electric current that can split water and produce hydrogen. This technology is particularly promising because it is far more efficient than the water-splitting devices currently in use. There's just one problem: price. As is the case with most new technologies, until they are produced and used frequently, they are far too expensive to be practical.

Storage Research

Hydrogen is currently stored primarily as a compressed gas or as a liquid at extremely low temperatures. The problem with these storage methods is that they often take up too much space. As a result, scientists are searching for ways to store larger amounts of hydrogen in smaller spaces at higher temperatures.

Syed Husain, CEO and president of QUANTUM Technologies, stands against a board showing a cutaway section of a tank designed to store compressed hydrogen gas. Mr. Husain announced a partnership deal with General Motors for the development of hydrogen storage devices that could make future fuel-cell electric vehicles competitive with gasoline cars.

Safety

Research into better and stronger containment systems is high on the list of energy priorities. So is the development of sensors that can detect hydrogen leaks. Because it's impossible to see or smell hydrogen gas, sensors will be the only way to tell if there is a leak. This will be an important safety feature on future hydrogen-powered vehicles.

THE BIG PICTURE

The U.S. Department of Energy Hydrogen Program's goal is to replace between two and four quads of conventional energy per year with hydrogen by the year 2010, and ten quads per year by 2030. One quad is equal to the amount of energy used by one million households.

In the very near future, there will be abundant supplies of hydrogen made from both renewable and nonrenewable sources. Later, as hydrogen technology improves, the world will decrease its dependence on fossil fuels. Virtually all hydrogen will be pollution free, made, very likely, from clean alternative energy sources like solar power. Eventually hydrogen may be made from other sources as well. One can never overestimate the miracles of science, or the minds of scientists. New technologies we can only dream about today will become commonplace tomorrow and new technologies will make visions of a world powered by hydrogen a reality. A hydrogen-based planet is just around the corner. Are you ready?

GLOSSARY

acid rain Rain that has become poisonous because of certain chemicals released into the air by burning fossil fuels.

carbon dioxide A gas released into the atmosphere with the burning of fossil fuels and through the normal breathing process of animals. A limited amount of carbon dioxide is essential for life.

compound A substance formed by combining two or more ingredients.

element A substance that cannot be broken down into other substances through chemical reactions.

emissions Substances released into the air.

energy The capacity for doing work.

fossil fuel A fuel, including coal, oil, or natural gas, formed millions of years ago in the earth from the remains of plants and animals.

fuel cell A device that changes the chemical energy of a fuel like hydrogen into electrical energy.

gaseous Having the form of, or being, a gas.

global warming The heating of the earth from natural and unnatural causes.

greenhouse effect The warming of the earth as a result of certain gases, especially carbon dioxide and methane, that trap heat in the lower atmosphere.

hydride A special type of compound containing hydrogen.

hydrogen The simplest, lightest, and most abundant of all the elements.

methane A gas that is produced when organic material decomposes as well as when coal is burned.

molecule A small particle of a substance.

nitrous oxide A polluting gas made of nitrogen and oxygen.

nonrenewable energy Energy from sources that can be used only once.

ozone A form of oxygen that occurs naturally in the stratosphere and prevents ultraviolet and other radiation from reaching the earth's surface. Ozone, when found near the earth's surface, however, causes respiratory problems in humans and animals.

photoconversion A process by which sunlight or electric currents are used to split water into hydrogen and oxygen.

politician A person who is elected to a job in government.

pollutant Something that puts dirt or poison into the air, ground, or water.

renewable energy Energy from sources that are replenished as they are used so that they never run out.

smog A low haze in the air that can be seen in many urban areas. Smog is caused by air pollutants.

steam reforming A method used to extract hydrogen from compounds that contain hydrogen.

water electrolysis A process that creates chemical changes when an electric current passes through water.

FOR MORE INFORMATION

American Hydrogen Association
1739 West 7th Avenue
Mesa, AZ 85202-1906
(480) 827-7915
Web site: http://www.clean-air.org

Hydrogen Energy Center
Portland, ME
(207) 831-8137
e-mail: info@h2eco.org
Web site: http://www.h2eco.org

Independent Power Producers' Society of Ontario
P.O. Box 1084, Station F
Toronto, ON M4Y 2T7
Canada
(416) 322-6549

International Association of Hydrogen Energy
P.O. Box 248266
Coral Gables, FL 33124
Web site: http://www.iahe.org

National Hydrogen Association
1800 M Street NW, Suite 300
Washington, DC 20036-5802
(202) 223-5547
Web site: http://www.hydrogenus.com

Natural Resources Canada
Office of Energy Efficiency
580 Booth Street, 18th Floor
Ottawa, ON K1A OE4
Canada
e-mail: general.oee@nrcan.gc.ca
Web site: http://oee.nrcan.gc.ca/english

United States Department of Energy
Energy Efficiency and Renewable Energy Network
1000 Independence Avenue SW
Washington, DC 20585
Web site: http://www.eren.doe.gov/RE/bioenergy.html

Web Sites

Due to the changing nature of Internet links, the Rosen Publishing Group, Inc., has developed an online list of Web sites related to the subject of this book. This site is updated regularly. Please use this link to access the list:

http://www.rosenlinks.com/lfe/hydr/

FOR FURTHER READING

Brown, Paul. *Energy and Resources* (Living for the Future). New York: Franklin Watts, 1998.

Challoner, Jack. *Eyewitness: Energy.* London: DK Publishing, 2000.

Chandler, Gary. *Alternative Energy Sources.* New York: Twenty-First Century Books, 1996.

Hoffmann, Peter. *Tomorrow's Energy*. Cambridge, MA: MIT Press, 2001.

Parker, Steve. *Earth's Resources* (Science Fact Files). Austin, TX: Raintree Steck-Vaughn, 2001.

Silverstein, Alvin. *Energy* (Science Concepts). New York: Twenty-First Century Books, 1998.

BIBLIOGRAPHY

Dunn, Seth. *Hydrogen Futures: Toward a Sustainable Energy System* (World Watch Paper 157). Washington, DC: Worldwatch Institute, 2001.

Padro, Catherine, E. Gregoire, and Francis Lau, eds. *Advances in Hydrogen Energy*. New York: Kluwer Academic/Plenum Publishers, 2000.

Rocks, Lawrence. *Fuels for Tomorrow*. Tulsa, OK: Penn Well Books, 1980.

Veziroglu, T. Nejat, ed. *Hydrogen Energy and Power Generation*. New York: Nova Science Publishers, 1991.

INDEX

CREDITS

About the Author

Chris Hayhurst works as a writer from his home in Colorado.

Photo Credits

Cover, pp. 22–23 © Roger Ressmeyer/Corbis; pp. 4–5 © Mark L. Stephenson/Corbis; p. 7 (left) © Farrell Grehan/ Corbis; p. 7 (right) Erik Schaffer/Ecoscene/Corbis; pp. 10–11 © Ken Eward/Photo Researchers, Inc.; p. 12 © Scott Camazine/Photo Researchers, Inc.; p. 15 © Lester V. Bergman/Corbis; pp. 17, 53 © AFP/Corbis; pp. 20, 26, 32, 33, 37, 38–39, 41, 44, 46–47, 48 © AP/Wide World Photo; pp. 24, 25 © Hulton/Archive/Getty Images, Inc.; p. 27 © NASA/Roger Ressmeyer/Corbis; pp. 30–31 © Reuters NewMedia, Inc./Corbis; p. 34 © Gary Braasch/Corbis.

Layout and Design

Thomas Forget